Unlock Your Potential: Using Data to Fuel Your Athletic Journey

Rony

TABLE OF COTENTS

CHAPTER 1. INTRODUCTION ....

    Quantifying Internal Load..

    Quantifying External Load ....

    Responses and Adaptations...

    Biochemical Markers ....

    Conclusion ....

CHAPTER 2. REVIEW OF RELATED LITERATURE.. 20

    Muscle Damage and the Inflammatory Response ....

    Resistance Exercise Skeletal Muscle Damage and the Inflammatory Response..............

    Common Biochemical Markers of Damage and Inflammation...

        C-Reactive Protein ....

        Interleukins and TNF-α...

Creatine Kinase ....

Cell Free DNA ....

Conclusion ....

CHAPTER 3. THE RELATIONSHIP BETWEEN CELL FREE DNA, CREATINE KINASE, C-REACTIVE PROTIEN, AND DELAYED ONSET MUSCLE SORENSS 48

INTRODUCTION ....

METHODS ....

Experimental Approach to the Problem ....

Participants ....

Training Protocol ....

Biochemistry ....

Performance Variables ....

Assessment of Soreness ....

Statistical Analysis ....

RESULTS ....

DISCUSSION ....

CONCLUSION ....

CHAPTER 4. THE RELATIONSHIP BETWEEN CELL FREE DNA AND VOLUME LOAD
.... 69

INTRODUCTION ....

METHODS ....

Experimental Approach to the Problem...

Participants...

One Repetition Back Squat Assessment ....

Resistance Training Session ....

Biochemistry ....

Statistical Analysis..

CHAPTER 5. SUMMARY AND FUTURE DIRECTIONS...

CHAPTER 1. INTRODUCTION

Sport Science is becoming increasingly more popular, and with it, the use of various athlete monitoring strategies. Monitoring has been demonstrated to be an asset in tracking athlete's well-being and performance across multiple sports. Sport and exercise training (practice, aerobic training, resistance training, competition) results in physical and psychological stress that can manifest as acute fatigue, inflammation, and muscle damage, possibly resulting in overreaching and overtraining. When planned appropriately, overreaching generally consists of a programmed period of time where athletes' complete intensified training before a competition (Hug, Mullis, Vogt, Ventura, & Hoppeler, 2003). If overreaching is unplanned or planned inappropriately, or training related fatigue is mismanaged, an athlete may experience overtraining. Overtraining is characterized by increased fatigue, decreased performance, higher psychological stress and risk for depression, and a compromised immune system (Dantzer, O'Connor, Freund, Johnson, & Kelley, 2008; Hug et al., 2003; Lee et al., 2017). Overtraining can manifest into physiological and biomechanical stress, both of which can be detected using internal and external load monitoring methods (Vanrenterghem, Nedergaard, Robinson, & Drust, 2017). The tools used to monitor internal load include heart rate (HR), session rate of perceived exertion (sRPE), surveys, contact forces, and soreness, while external load can be quantified using distance covered and accelerations and decelerations (Vanrenterghem et al., 2017). Responses to training load can also be used in athlete monitoring, with VO_{2max}, gait asymmetries, sprint time, and jump monitoring,

Quantifying Internal Load

Monitoring HR is a common means of tracking internal load in athletes, due to HRs linear relationship with the rate of oxygen consumption during exercise (Halson, 2014). HR can also be used to calculate training impulse (TRIMP) and its derivatives. TRIMP was originally conceived by Banister and colleagues by combining an athlete's heart rate response with exercise duration, but this method is limited by the requirement of steady-state-rate (Borresen & Lambert, 2008). Additional methods have been outlined in the literature that use three heart rate zones determined by an athlete's lactate threshold (Halson, 2014). Time spent is each zone is then multiplied by a weighting factor and summed to calculate TRIMP (Halson, 2014).

While HR monitoring methods may be suggestive of an athlete's physiological training load, it may not be indicative of mood disturbances associated with elevated fatigue (Armstrong & Vanheest, 2002). The use of subjective questionnaires has proven useful when monitoring an athlete's well-being and fatigue levels when compared to physiological measures, with some literature suggesting subjective measures are better correlated with well-being (Saw, Main, & Gastin, 2016; Vanrenterghem et al., 2017). There are multiple questionnaires available for athlete monitoring; the Profile Mood State (POMS), Recovery Stress Questionnaire for Athletes (RESTQ-S) and the Daily Analyses of Life Demands of Athletes (DALDA) are commonly used options reported in the literature (Saw et al., 2016). Questionnaires correlate with performance decreases corresponding with increase training loads and are useful instruments to identify risks for depressive symptoms that may manifest from high fatigue and overtraining (Coutts, Reaburn, Piva, & Rowsell, 2007; Grant, Van Rensburg, Collins, Wood, & Du Toit, 2012).

RPE is a subjective measure that can supplement the planning of practice and training sessions, helping to ensure athletes are completing the prescribed workload (Coutts, Reaburn, Murphy, Pine, & Impellizzeri, 2003; Impellizzeri, Rampinini, Coutts, Sassi, & Marcora, 2004). RPE has a moderate to high correlation with player load measured by global positioning systems (GPS) (Rossi et al., 2017; Vanrenterghem et al., 2017). sRPE is a modified version of the RPE scale of measurement that is more direct measurement of internal training load (Halson, 2014). sRPE is calculated by multiplying athlete reported RPE by session duration (Gaudino et al., 2015). sRPE is strongly correlated with $\%HR_{peak}$, $\%HR_{reserve}$, distance covered, running speed, and player load during intermittent team sports (Herman, Foster, Maher, Mikat, & Porcari, 2006; Scott, Black, Quinn, & Coutts, 2013). RPE and sRPE are cost effective methods for athlete monitoring and improve real-time adjustments to a training program.

As with physiological monitoring methods, subjective tools have their disadvantages. The subjective nature relies on the athlete's honesty and ability to understand their current fatigue levels to be valid (Herman et al., 2006; Nässi, Ferrauti, Meyer, Pfeiffer, & Kellmann, 2017; Saw, Main, & Gastin, 2015). Repeated long term use of surveys may result in monotony and decreased compliance on the part of the athlete, causing the results to be misleading (Nässi et al., 2017). This disadvantage should not discourage the use of surveys in monitoring, due to their cost effeteness, but it may be pertinent to include additional monitoring modalities to compliment subjective measures.

Quantifying External Load

The use of wearable devices for monitoring internal training load has been gaining popularity (Ehrmann, Duncan, Sindhusake, Franzsen, & Greene, 2016). Global Position Systems (GPS) and accelerometers are two popular wearable micro sensors capable of indicating the volume of work completed by an athlete. The use of GPS has been shown to be valid and reliable method when quantifying training load in running based sports (Vanrenterghem et al., 2017). It has become popular in sports such as soccer and has helped decrease injury risk due to overtraining (Ehrmann et al., 2016). One negative aspect of GPS is that it is unable to quantify accelerations and decelerations experienced by an athlete during practice or competition. These

variables require greater energy use and cause more fatigue than running at a constant velocity
(Di Prampero et al., 2005). Accelerometers have been used in combination with GPS to quantify
distance as well as accelerations and decelerations (Vanrenterghem et al., 2017). Knowing the
distance and accelerations completed by an athlete allows coaches to decide when to make
substitutions, allowing players who have generated more work to rest (Gaudino et al., 2013).
Wearable technology with accelerometers and gyroscopes, such as inertial measurements units,
are continuing to improve and have shown promise when compared to the gold standards of
force plate and motion capture systems (Benson et al., 2020; Clemens et al., 2020; Gómez-
Carmona, Bastida-Castillo, González-Custodio, Olcina, & Pino-Ortega, 2020).

Responses and Adaptations

Monitoring acute changes in performance tests due to training load variations provides
information regarding an athlete's current fatigue state. Chronic adaptations to training load
should elicit performance improvements such as increases to VO_{2max} seen in endurance athletes
and greater jump height in strength-power and team sport athletes.

VO_{2max} is an important factor contributing to success in endurance sports, with athletes at
elite levels able to achieve greater VO_{2max} values compared to their intermediate and amateur
counterparts (Esco, Flatt, & Nakamura, 2016), with increases in VO_{2max} demonstrated in
successful endurance training programs (Farsani & Rezaeimanesh, 2011). While changes in
VO_{2max} can be used to track performance variations, the VO_{2max} testing is fatiguing and requires
special instrumentation normally exclusive to laboratory settings (Coates, Hammond, & Burr,
2018; Coetzer et al., 1993; Gabryś et al., 2019). These factors have led to the development of
field tests that can estimate aerobic power, such as the beep test and yo-yo intermittent test, both

of which have been validated as appropriate measure to estimate VO_{2max} (Dourado et al., 2001; Gabryś et al., 2019; Metaxas, Koutlianos, Kouidi, & Deligiannis, 2005).

Ground reaction forces (GRF) can be monitiered to help track gait alterations due to fatigue. Decreased locomotor function and impaired movement kinematics may arise from acute and chronic fatigue, manifesting from altered and/or assymetric variations in GRF, which increases injury risk (Bellenger, Arnold, Buckley, Thewlis, & Fuller, 2019; Strohrmann, Harms, Kappeler-Setz, & Troster, 2012). Tracking running gait or vertical jump asymmetry using force plates or ground contact time differences with motion capture systems have both been shown to be valid and reliable (McGrath, Greene, O'Donovan, & Caulfield, 2012).

A variety of performance tests can be used in athlete monitoring. Two common methods are tracking in-game performance and changes to physical abilities such as vertical jump, change of direction, and sprint time. Decreased performance for an individual athlete during a game has been correlated with an increased risk of overtraining and injury (Urhausen & Kindermann, 2002). Quantifying performance variations is useful for coaches to get a performance baseline established for an athlete and know when to complete substitutions if they are underperforming but does not allow for training modifications prior to a game to ensure an athlete's readiness. Non-fatiguing skill-based tests that are related to in-game performance can be better utilized to monitor the fatigue and wellness of an athlete. Decreases in vertical jump height and increases in sprint time both correlate with neuromuscular fatigue and reduced game performance across multiple sports (Haugen & Buchheit, 2016; Loturco et al., 2017; Sams, Sato, DeWeese, Sayers, & Stone, 2018). Vertical jump is considered the most popular method of monitoring neuromuscular fatigue (Sams et al., 2018), with multiple modalities used to collect data. Force plates, Vertec, wearable devices, and 3D motion capture are all viable options (Leard et al.,

2007; MacDonald, Bahr, Baltich, Whittaker, & Meeuwisse, 2017), with force plates and 3d motion capture cited as the gold standards (Leard et al., 2007; Loturco et al., 2017). Vertical jump can also be used to track training adaptations, with increases in jump height seen with strength and power increases (Bazyler et al., 2018). Athlete monitoring using skill-based tests provides insight regarding fatigue levels of an athlete but lacks direct information on the state of their immune system and physiological damage sustained from training (Coutts et al., 2007).

Biochemical Markers

A single training session affects multiple biochemical markers, while multiple sessions may exacerbate these responses. Due to the vast number of different biochemical markers used, this paper will be limited to creatine kinase (CK), c-reactive protein (CRP), interleukins 1 (IL-1) and 6 (IL-6), tumor necrosis factor alpha (TNF-α), and cell free DNA (cf-DNA).

CK is an enzyme found in skeletal muscle that catalyzes the phosphorylation of creatine by adenosine triphosphate, producing phosphocreatine (McLeish & Kenyon, 2005). Exercise induced skeletal muscle damage results in elevated circulating CK levels and triggers the immune response, initiating the healing process (Raeder et al., 2016; Soneja, Drews, & Malinski, 2005). CRP is an acute phase response plasma protein that has an important role in the inflammatory process through the activation of the compliment pathway and stimulation phagocytosis (Black, Kushner, & Samols, 2004; Neumaier, Metak, & Scherer, 2006). The relationship CK and CRP have skeletal muscle damage and fatigue has resulted in their usage as markers for athlete monitoring (Brancaccio, Maffulli, & Limongelli, 2007; Pyne, 1994; Raeder et al., 2016). The acute phase response is regulated by cytokines such as interleukins and TNF-α (Bansal, Pandey, Deepa, & Asthana, 2014).

17

Interleukins and TNF-α are cytokines that, when bound to their respective receptors,
signal communication among leukocytes (Akdis et al., 2011; Victor & Gottlieb, 2002). Exercise
tends to elevate levels of IL-1, IL-6, and TNF-α (Pyne, 1994; Smith et al., 2000), making them
another popular group of biomarkers used in athlete monitoring (Ostrowski, Rohde, Asp,
Schjerling, & Pedersen, 1999; Vaisberg, de Mello, Seelaender, dos Santos, & Rosa, 2007). The
inflammatory response that involves the aforementioned cytokines may be reflected by
circulating levels of cf-DNA (Frank, 2016).

cf-DNA is a marker that is commonly used in the medical field to monitor patient status
in conditions including cancer (Volik, Alcaide, Morin, & Collins, 2016) type II diabetes (Bryk et
al., 2019), and heart failure (Yokokawa et al., 2019). The mechanisms are not entirely
understood but with neutrophil extracellular traps are thought to be the primary source due to
exercise (Beiter, Fragasso, Hudemann, Nieß, & Simon, 2011; Swystun, Mukherjee, & Liaw,
2011; Wilson et al., 2017), with apoptosis and necrosis as secondary sources (Schwarzenbach,
Hoon, & Pantel, 2011). The use of cf-DNA in athlete monitoring is in its infancy, but shows
promise due to short half life and high sensitivity (Fatouros et al., 2006; Tug et al., 2017).

Conclusion

While in-game performance is the most valid form of monitoring, elevated fatigue in one
or more athletes may have a negative impact on competition outcome. (Saw et al., 2016).
Monitoring throughout the year can help ensure athletes are in an optimal state of readiness
during key competition periods. An effective monitoring program includes both internal and
external modes of monitoring (Gaudino et al., 2013). cf-DNA shows promise as a biochemical
marker for athlete monitoring due to its short half-life and ability to elucidate immune status
(Fatouros et al., 2006; Tug et al., 2017). Further research is required on cf-DNA to better

understand its kinetics during and after resistance training, and to determine how sensitive cf-

DNA is to changes in volume-load.

CHAPTER 2. REVIEW OF RELATED LITERATURE

Skeletal muscle is composed of cells that are resistant to cell turnover and injury. Death of these cells occurs infrequently under circumstances such as toxic stimuli, degenerative diseases, ischemia, genetic conditions such as Duchenne and Becker Muscular Dystrophies, mechanical injury, and physical exercise (de Souza, & Gottfried, 2013). Injury within the muscle can occur to the connective tissue that anchor muscle to bone and transmit force, or within the myofibril to the contractile filaments responsible for force generation (Kääriäinen, Järvinen, Järvinen, Rantanen, & Kalimo, 2000). This can result in decreased muscle function, lower force production, and increase muscle degradation over time (Clarkson & Newham, 1995; Tabebordbar, Wang, & Wagers, 2013). Indicators of muscle damage are important monitoring tools for both clinicians and practitioners in sport. For instance, creatine kinase (CK) is an enzyme found in skeletal muscle, cardiac muscle and brain cells that catalyzes the phosphorylation of creatine by adenosine triphosphate (McLeish & Kenyon, 2005). CK is commonly used to monitor muscle damage and necrosis resulting from, but not limited to, myocardial infarctions (Kost, Kirk, & Omand, 1998), strokes (Ay, Arsava, & Sarıbas, 2002), sport related activities (Lazarim, 2009) and resistance training (Brancaccio, Maffulli, & Limongelli, 2007). Elevated levels of CK in circulating blood is associated with greater muscle damage (Horta, Bara Filho, Coimbra, Miranda, & Werneck, 2017) and this damage may compromise qualities important for exercise and sport performance such as strength, power, and range of motion (Peake, Neubauer, Della Gatta, & Nosaka, 2017). Therefore, CK and other biochemical markers, may be valuable indicators for use in an athlete monitoring program to quantify muscle damage, athlete status, and the inflammatory response precipitated by muscle damage.

The damage sustained by the muscle initiates the inflammatory response and the process of repair (Baoge et al., 2012). Inflammation can be localized, in cases such as blunt trauma (Feng et al., 2007), or systemic as seen with cancer, disease, or some exercise (Calle & Fernandez, 2010; Davies & Hagen, 1997; Libby, 2007). The degree of inflammation can provide valuable information regarding the magnitude and nature of the stressor. Inflammation provides insight into the status of a patient or athlete, with higher inflammatory levels associated with a weakened immune system, resulting in higher disease risk among all populations (Dantzer, O'Connor, Freund, Johnson, & Kelley, 2008) and decreased performance capabilities demonstrated in athletes (Buonocore, Negro, Arcelli, & Marzatico, 2015). These concerns have led to the monitoring of inflammation using physiological markers, similar to monitoring muscle damage using CK. The capacity of the immune system to repair damage sustained from exercise is limited, with strong evidence outlining the concerns of overtraining (Lee et al., 2017), demonstrating the need to monitor training status. Numerous physiological markers have been used to measure the inflammatory response. C-reactive protein (CRP) is an acute-phase protein playing multiple roles in inflammation. Plasma concentrations increase during numerous inflammatory states such as cancer (Allin, Bojesen, & Nordestgaard, 2009), type II diabetes (Arnalich et al., 2000), cardiovascular disease (Brull et al., 2003), and resistance training (Libardi, De Souza, Cavaglieri, Madruga, & Chacon-Mikahil, 2012), leading to the use of CRP in the medical field for patient monitoring (Black, Kushner, & Samols, 2004). Cell signaling proteins such as cytokines including a variety of interleukins and tumor necrosis factor alpha (TNF-α) are also common indicators of inflammation. While the nature of their actions differ, some cytokines are pro-inflammatory signalers (TNF-α, interleukin-1 (IL-1)) (Keller, Rüegg, Werner, & Beer, 2008; Vasanthi, Nalini, & Rajasekhar 2007), others have anti-inflammatory actions (interleukin 10 (IL-10)) (Cao, Zhang, Edwards, & Mosser, 2006), and in some cases have both pro- and anti-inflammatory affects (interleukin 6 (IL-6)) (Rose-John, 2012). Cell free DNA (cf-DNA), a novel marker in exercise and sport science, has also demonstrated promise as an indicator of tissue damage and inflammation (Chang et al., 2003; Gentles et al., 2017; Spindler et al., 2016), but additional research is still required to assess its utility and origins in circulation (Fatouros et al., 2006).

This literature review will outline the inflammatory response and the roles of specific inflammatory markers including, c-reactive protein (CRP), interleukins, tumor necrosis factor alpha (TNF-α), creatine kinase (CK), and cell free DNA (cf-DNA). This review will also describe how these markers are related to exercise.

Muscle Damage and the Inflammatory Response

Damage to skeletal muscle can occur via multiple mechanisms including acute damage, such as blunt force trauma and extreme temperatures, contraction-induced damage, and inflammatory myopathy. These mechanisms can lead to mild disruption of the sarcomeres or myofibril necrosis, in more extreme cases (Tabebordbar et al., 2013). Disruption may occur to multiple sites within the sarcomere depending on the strength, type, and duration of the stimulus. Transverse tubules (t-tubules) surrounding the sarcomere allow for the transport of calcium to the contractile filaments. The t-tubules can be forced out of position when the muscle is placed under stress, which may inhibit calcium dynamics and decrease muscle function (Cooper & Head, 2015). Within the sarcomere, damage to the costamere region may occur from stimuli including exercise (Bloch et al., 2002), blunt force trauma (Kami et al., 1993), or muscular dystrophy (Cooper & Head, 2015). This may result in decreased capacity for force transduction from the contractile filaments to the surrounding tissue and disruption to the shape of the sarcomere (Ervasti, 2003).

Similar to the costamere, the cytoskeleton is a network of proteins, which interact to stabilize the sarcomere and are responsible for muscle contraction. Key cytoskeleton proteins included titin, desmin, actin, and myosin (Greaser, 1991). The contractile filaments may sustain damage when exposed to a stressor. Titin anchors the myosin filaments to the z-line and contributes to the resting tension of stretched muscle fibers (Magid & Law, 1985). The attachments of titin to myosin can become uncoupled when the sarcomere is under extreme stretching or radiation (Morgan & Allen, 1999). Free titin will be recruited to replace the damaged molecules, but the pool is limited, possibly resulting in deformed sarcomeres with decreased contraction and stretch resistant capabilities (Morgan & Allen, 1999). Desmin also plays a role in sarcomere integrity through z-lines, connecting adjacent z-disks and anchoring the apex and base to the costamere (Morgan & Allen, 1999). This allows desmin to assist in the alignment of z-disks across fibers and the transmission of lateral tension to the costamere (Morgan & Allen, 1999). When desmin is compromised, this may diminish the structural integrity of the myofibril (Milner, Weitzer, Tran, Bradley, & Capetanaki, 1996). Substantial losses of desmin in mouse models have resulted in tissue fibrosis and subsequent mechanical stiffness (Chapman et al., 2014).

The sarcolemma surrounding a myofibril can become perturbed, increasing membrane permeability for hydrophilic substances that inhibit membrane function (Rocha, Leonardo, De Souza, Palma, & Da Cruz-Höfling, 2008). Increased permeability allows damaged intramuscular proteins to leak into circulating blood, decreasing muscle protein content (Fisher, Baracos, Shnitka, Mendryk, & Reid, 1990). Elevated blood protein content is one factor that causes an immune and subsequent inflammatory response, which promotes clearing the debris from the damaged tissue (Mercier, Breuille, Mosoni, Obled, & Patureau Mirand, 2002).

The body's inflammatory response to injury or disease, promotes healing and repair of the affected tissues (Kim, West, & Byzova, 2013) and is the first step in the three-phase process of healing: inflammation, tissue formation, and tissue remodeling (Eming, Krieg, & Davidson, 2007). Damage to tissues induces the inflammatory response and the first step in the healing process (Chen et al., 2018). Vasodilation is a staple of inflammation and occurs through the signaling of nitric oxide (NO), prostacyclin (PGI_2), prostaglandin E_1 (PGE_1) and E_2 (PGE_2), and prostaglandin D_2 (PGD_2) (Ialenti, Ianaro, Moncada, & Di Rosa, 1992). Increased blood flow to damaged tissues allows for the healing process to begin (Ialenti et al., 1992; Guzik, Korbut, & Adamek-Guzik, 2003), by delivering more oxygen and nutrients to damaged tissue and increasing the available leukocyte count for fighting infection or the clearing of debris (Ryan & Majno, 1977). The first leukocytes present during inflammation are typically neutrophils, which are recruited by proinflammatory cytokines from tissue resident sentinel leukocytes (Kolaczkowska & Kubes, 2013). Neutrophils can eliminate pathogens and clear debris via multiple mechanisms and the ability of neutrophils to phagocytize other molecules has long been understood. Neutrophils encapsulate target molecules and kill or catabolize those molecules using reaction oxygen species (ROS) (Kolaczkowska & Kubes, 2013). More recently discovered, neutrophils can also engulf target molecules via neutrophil extracellular traps (NETs). NETs are composed of DNA with attached proteins and enzymes, some able to produce ROS (Kolaczkowska & Kubes, 2013). Once released, NETs are capable of trapping pathogens and debris, preventing continued circulation of these potentially harmful substances (Kolaczkowska & Kubes, 2013). The ROS released from NETs may be similar to their actions during phagocytosis; the trapped materials are catabolized (Kolaczkowska & Kubes, 2013). Neutrophils will create a positive feedback loop through the release of proinflammatory cytokines, recruiting more neutrophils to the area (Xing & Remick, 2003). Once neutrophil stimulation has ended, macrophages mobilize to the site and begin phagocytizing neutrophils (Eming et al., 2007). Like neutrophils, macrophages release pro-inflammatory cytokines which further promote macrophage activity; these pro-inflammatory cytokines include interleukin 1 beta (IL-1β), interleukin 6 (IL-6), and TNF -α (Jovanovic et al., 1998; Xagorari et al., 2001). Once debris and/or pathogens have been phagocytized, macrophages shift and begin releasing the anti-inflammatory cytokine interleukin-10 (IL-10), initiating the next stages of the healing process (Cao et al., 2006).

Macrophages also play an important role in the second stage of the healing process, tissue formation. Macrophages synthesize growth factors such as tissue growth factors beta and alpha, platelet-derived growth factor, and basic fibroblast growth factor (Eming et al., 2007). Epithelial cells cover the injury site, termed "Epithelialization", and new granulation tissue is formed (Pastar et al., 2014). These newly formed tissues, paired with fibroblasts and leukocytes, begin to cover the tissue, transitioning into phase three, tissue remodeling (Pastar et al., 2014). The inflammatory response to damage is necessary for tissue repair, but inflammation may also be associated with acute and chronic negative effects. For instance, the immune system may be compromised while inflammation is acutely elevated (Dantzer et al., 2008) and conditions such as type II diabetes (Pickup, 2004) and cancer (Franceschi & Campisi, 2014; Marx, 2004) may result in a chronic inflammatory state that vitiates the transition into phase two and three of the healing process. Considering the host of negative influences that inflammation may encourage, it may be valuable for practitioners to monitor the inflammatory state of clinical and athletic populations. One activity that both the general public and athletes engage in, is resistance training. Resistance training, which will largely be the focus of the investigations included in this dissertation, can cause a significant inflammatory/immune response which deserves continued investigation.

Resistance Exercise Skeletal Muscle Damage and the Inflammatory Response

Resistance training can result in skeletal muscle damage in individuals of all fitness levels, with lesser trained individuals experiencing greater damage when engaged in similar relative intensities (Pyne, 1994). This damage may cause soreness generally referred to as delayed onset muscle soreness (DOMS) (Hackney, Engels, & Gretebeck, 2008). The damage caused as a result of resistance training is positively associated with higher training volumes and intensities, placing greater stress on skeletal muscle causing increased fatigue (Peake et al; Pyne, 1994). Contraction induced damage results from the overextension of sarcomeres, which is most pronounced during the eccentric phase (Tabebordbar et al., 2013). Z-line streaming, disorganization of myofilaments, and T-tubule damage have all been associated with contraction induced damage (Clarkson & Hubal, 2002). T-tubule damage can increase intracellular calcium which promotes the release of pro-inflammatory cytokines and ROS that may cause additional damage of sarcomere proteins and sarcolemma, and potentially induce fiber necrosis (Tabebordbar et al., 2013; Yu, Liu, Carlsson, Thornell, & Stål, 2013). Sarcolemma damage and necrosis are only seen in extreme cases of voluntary contraction induced damage and is primarily exclusive to training with an eccentric focus (Yu et al., 2013).

The body responds to the breakdown of skeletal muscle by initiating the inflammatory response. Exercise training causes an acute-response systemic inflammation that differs from trauma and disease and has been demonstrated to be beneficial to immune competence

(Fehrenbach & Schneider, 2006). The inflammatory response can provide valuable information regarding the magnitude and nature of the training session. A dose-response relationship between the intensity and/or duration of exercise and inflammatory activity has been outlined in the literature, where it has been shown that increased exercise volume and intensity enhances performance related adaptations (Gormley et al., 2008; Kramer et al., 1997). The inflammatory response, followed by recovery, are important promoters of exercise related adaptations such as strength and hypertrophy.

As discussed previously, NO is largely responsible for vasodilation which promotes the delivery of leukocytes to muscle, thereby enhancing the healing process. Increased NO and macrophage activity are associated with greater damage to the skeletal muscle after resistance training (Freidenreich & Volek, 2012; Macedo et al., 2016). NO and macrophages also play a critical role in the stimulation of satellite cells (Bloomer, Williams, Canale, Farney, & Kabir, 2010; Freidenreich & Volek, 2012). Satellite cells are muscle stem cells located between the sarcolemma and basal lamina of a muscle fiber which aid in muscle recovery, growth, and remodeling (Herman-Montemayor, Hikida, & Staron, 2015). Stem cells donate nuclei to muscle fibers, as well as create precursor cells, termed myoblasts, which fuse to existing cells providing needed agents for recovery (Dangott, Schultz, & Mozdziak, 2000; Merly, Lescaudron, Rouaud, Crossin, & Gardahaut, 1999; Tabebordbar et al., 2013). The process of muscle derived nitric oxide as a signaler for satellite cells is regulated by the cyclooxygenase enzymatic pathway, specifically cyclooxygenase-2 (COX-2). COX-2 upregulates cell proliferation and growth, possibly by increasing certain prostaglandins thought to increase DNA synthesis and cell proliferation (Cao & Prescott, 2002). It has been shown that anti-inflammatory drugs which inhibit inducible NO synthase (iNOS), the enzyme responsible for the synthesis of NO, downregulate the COX-2 pathway (Chien et al., 2018). The downregulation of the COX-2 pathway is thought to diminish the satellite cell response to muscle damage and the inhibitory effect of COX-2 is thought to decrease the hypertrophic response caused by long term resistance training, as well as the inflammatory response of the prostaglandins (Schoenfeld, 2012). Reduced prostaglandin production results in less blood and nutrient delivery to the damaged muscle (Aronoff, Oates, & Boutaud, 2006), suggesting that hypertrophy, and potentially strength resulting from less hypertrophy, is suppressed when muscle inflammation fails to occur and the COX-2 pathway is down-regulated (Schoenfeld, 2012).

High volume resistance training and resistance training that focuses on or includes eccentric contractions, has been shown to decrease acute performance (Carroll et al., 2017). This acute decrease in performance is generally not a concern for most athletes, unless improperly timed prior to a competition, and is part of the normal training process. As eluded to previously, the acute inflammatory response associated with muscle damage causes beneficial skeletal muscle remodeling that may have a positive impact on strength and hypertrophy (Wilborn, Taylor, Greenwood, Kreider, & Willoughby, 2009). While acute periods of elevated inflammation promote adaptations, assurances should be made that sufficient recovery occurs between sessions, including proper sleep management.

Sleep is of importance for recovery from a variety of stressors, including resistance training and other sport related activities (Fullagar et al., 2015). Reduced sleep quantity and quality may delay recovery after sport related training and poor recovery may promote sleep disturbances (Fullagar et al., 2015; Lastella et al., 2018). Longer sleep duration has been shown to have a strong positive correlation with in-season performance, suggesting a dose-response effect of sleep on performance (Brandt, Bevilacqua, & Andrade, 2017; Juliff, Halson, Hebert,

Forsyth, & Peiffer, 2018). Sleep deprivation alone has been shown to increase circulating leukocytes, resting heart rate, blood pressure (Meier-Ewert et al., 2004) and energy requirements (Mullington, Haack, Toth, Serrador, & Meier-Ewert, 2009). If sleep deprivation continues unmitigated, it may increase disease risk, reduce glucose sensitivity, decrease performance, and lead to chronic inflammation (Meier-Ewert et al., 2004; Roth et al., 2017).

Chronic inflammation predisposes athletic population to increased injury and illness risk due to the compromised state of the immune system (Lee et al., 2017), and may negatively impact mood state, increasing the risk of depression (Dantzer et al., 2008). Chronic inflammation resulting from elevated training stress, inconsistent schedules and/or under-recovery, may lead to overtraining syndrome, a state of chronic fatigue which decreases performance (Lastella et al., 2018; Roth et al., 2017). The prevalence of overtraining among one population of elite athletes was shown to be 37%, suggesting that overtraining should be a concern for coaches and athletes (Fehrenbach & Schneider, 2006). The nature of an inflammatory response begins to change as inflammation becomes chronic. As detailed previously, macrophages are signaled to the injured areas to phagocytose the damaged tissue, as they would under normal circumstances (Jovanovic et al., 1998). These macrophages then release pro-inflammatory cytokines such as interleukin 1 beta (IL-1β), interleukin 6 (IL-6), and TNF -α, resulting in additional leukocytes and blood delivery to the damaged area (Jovanovic et al., 1998). When an athlete is engaged in a well-managed training program, adequate recovery is provided and their immune system is not compromised, macrophages will transition from releasing the pro-inflammatory cytokines above and begin releasing the anti-inflammatory cytokine interleukin-10 (IL-10), thereby initiating the next stages of the healing process (Cao et al., 2006). If an athlete's immune system is compromised due to chronic sleep deprivation, overtraining, or other sources of chronic stress, the transition from pro- to anti-inflammatory cytokine production is inhibited. This ultimately perpetuates inflammation and stunts the recovery process. Caution should be taken to ensure a single training session does not cause injury or excessive muscle damage. Extreme cases of acute highly fatiguing exercise can result in rhabdomyolysis, a condition where elevated amounts of skeletal muscle break down and the contents leak into the blood (Sauret, Marinides, & Wang, 2002). This results in high levels of inflammation and can cause kidney failure due to the vast amount of proteins that now must be processed by the renal system (Sauret et al., 2002).

The capacity of the immune system to repair damage sustained from exercise is limited, with strong evidence outlining the concerns of overtraining (Lee et al., 2017). The relationship between overtraining and disease risk demonstrates the need to monitor training status (Dantzer et al., 2008; Gleeson & Bishop, 2000). Physiological indicators of inflammation and muscle damage can be used to track an individual's immune response and may be indicative of fatigue associated with training (Lee et al., 2017). Numerous physiological markers have been used to measure muscle damage and the inflammatory response, including creatine kinase (CK) c-reactive protein (CRP), interleukins, tumor necrosis factor alpha (TNF-α), and cell free DNA (cf-DNA) (Fatouros et al., 2006; Lee et al., 2017)

Common Biochemical Markers of Damage and Inflammation

Numerous biochemical markers are used to quantify tissue damage, systemic inflammation and acute inflammatory responses. For instance, serum CRP is a common non-specific indicator of inflammation; specific interleukins and TNF-α are frequently used as well. CK is often measured in addition to inflammatory markers and used to assess the magnitude of muscle damage (Brancaccio et al., 2007). Circulating Cf-DNA is a marker commonly used when monitoring health status in patients with chronic conditions and infections (Elshimali, Khaddour,

Sarkissyan, Wu, & Vadgama, 2013). Elevated levels with exercise have been shown to be a predictor of fatigue and inflammation (Breitbach, Tug, & Simon, 2012). The use of multiple markers may be advantageous when assessing damage and/or inflammation, due to substantial between subject variability in initial baseline values and the response to the same stimuli (Lee et al., 2017).

C-Reactive Protein

CRP is an acute phase protein that is produced by hepatocytes (Punyadeera, & Slowey, 2013) and possibly human adipocytes (Calabro, Chang, Willerson, & Yeh, 2005). CRP has been shown to be associated with both anti and pro-inflammatory processes. The anti-inflammatory mechanisms include inducing the expression of interleukin-1 receptor antagonist (IL-1ra). IL-1ra can bind to the IL-1 receptor on cells, blocking IL-1 from binding, thereby impeding the inflammatory effects of IL-1 (Bresnihan et al., 1998). CRP can also increase the release of anti-inflammatory cytokine IL-10, and decrease the synthesis of interferon-γ, which mitigates the stimulation of macrophages (Schroder, Hertzog, Ravasi, & Hume, 2004; Black et al., 2004).

CRP is more commonly known for its pro-inflammatory functions. Common baseline CRP levels for healthy individuals are less than 2.0mg*L^{-1} in men and 2.5 mg*L^{-1} in women (Kostrzewa-Nowak et al., 2015). A CRP level of 10 mg*L^{-1} or greater is considered a clinically significant level of inflammation (Kostrzewa-Nowak et al., 2015). Levels of CRP have been shown to rise as much as 1,000-fold from baseline levels in response to trauma or infection (Kostrzewa-Nowak et al., 2015). These substantial increases make it valuable when qualifying the level of inflammation that has occurred to a patient, athlete or subject. A CRP level greater than 3.0 mg*L^{-1} is considered high risk (Kostrzewa-Nowak et al., 2015). It has also been demonstrated that baseline CRP declines as physical fitness, measured using the Bruce Protocol, improves (Aronson et al., 2004). This is valuable as research has shown that chronically elevated CRP is positively correlated with coronary disease and diabetes (Aronson et al., 2004).

CRP can increase the release of the inflammatory cytokines IL-1, IL-6, IL-8, IL-18, and TNFα, up-regulating adhesion molecule expression in endothelial cells, and activating the complement system (Black et al., 2004). Water retention increases are seen with elevated levels of CRP as well, which is a common trait seen with inflammation (Vicenté-Martínez et al., 2004). The complement system includes over 30 proteins and plays an importnat role in the body's defense and is closely linked to the inflammatory response (Sarma & Ward, 2011). Apoptosis, or programmed cell death, can be initiated by this system, which will result in increased monocyte delivery to the affected area through inflammation (Sarma & Ward, 2011). Adaptive immunity has also been shown to be a central responsibility of the complement system, which reduces instances of pathogen re-envision (Carroll, 2004). Additionally, the complement system plays a role in satellite cell proliferation through upregulation of tissue regeneration (Syriga & Mavroidis, 2013). The extent to which CRP increases the production of the inflammatory response has been shown to be dose dependent. As CRP levels increase, additional CRP binds to macrophages and stimulates the release of IL-1β, IL-6, and TNFα (Ballou, & Lozanski, 1992), which continues to stimulate the release of satellite cells from the basal lamina (Syriga & Mavroidis, 2013).

Baseline levels of CRP are used to monitor inflammatory status in a variety of populations including conditions such as type II diabetes and cancer (Allin et al., 2009; Festa, D'Agostino, Tracy, & Haffner, 2002), and non-clinical settings in athletes. While chronic levels of CRP in blood is inversely related to physical fitness, a single bout of exercise may cause CRP to increase acutely (Kostrzewa-Nowak et al., 2015). The magnitude of acute CRP response can be used as an indicator of the stress a subject or athlete has encountered. Elevated CRP levels have been seen in response to a soccer match and this response seems to similar between elite males to elite females when there is no significant difference in relative intensity (Souglis et al., 2015). Similar findings have been shown between young and middle-aged men when using a resistance training protocol, with no significant changes to CRP when relative intensity was held constant (Gordon III et al., 2017). This data illustrates that CRP can be used to reflect the relative intensity and subsequent inflammatory response associated with a soccer match or a resistance training session. The studies addressed above suggest that CRP can provide coaches with insight to the degree of inflammation that will result from training, and the time required for the inflammatory response to return to baseline (Gordon III et al., 2017). Athletes can sustain elevated CRP levels through competition. CRP levels may remain high, especially during training that includes excessive volumes and/or intensities (Petibois, Cazorla, Poortmans, & Déléris, 2002). Prolonged periods of elevated CRP without adequate rest may lead to overtraining (Petibois et al., 2002), which was previously addressed in this paper. The degree to which symptoms of overtraining, decreased performance, negative mood state, immune suppression, fatigue, and hormone fluctuations (Booth, Probert, Forbes-Ewan & Coad, 2006), can interfere with the ability to compete, outlines the importance of monitoring an athletes' recovery state. The severity of symptoms also depends on the degree of inflammation and fatigue (Booth et al., 2006).

The usefulness of quantifying inflammation levels has been demonstrated in research, as it provides insight into the magnitude of stress a training session may cause an athlete. CRP is a functional tool but has some limitations which make using it difficult in an athlete monitoring program. Baseline CRP levels vary substantially between individuals which may cause difficulty when interpreting changes or the results from previous studies (Costello et al., 2018; Marcell, McAuley, Traustadóttir, & Reaven, 2005. Additionally, previous studies have described the response time for CRP, which may remain elevated beyond 48 hours after a single training session (Gentles et al., 2017; Gordon III et al., 2017). The fact that CRP may require greater than 48 hours to return to baseline makes day to day monitoring difficult due to the high frequency of training most athletes engage in (Gordon III et al., 2017). The kinetics of CRP, in addition with its 19-hour half-life (Pepys & Hirschfield, 2003), decreases the practicality of using CRP to quantify inflammation from individual training sessions throughout the week.

Interleukins and TNF-α

Interleukins, a group of cytokines, are released by cells and have a specific function in between-cell communication (Zhang & An, 2007). Interleukins are produced primarily by leukocytes and have multiple functions including pro- and anti-inflammatory signaling and pathological pain (Zhang & An, 2007). TNF-α is also a cytokine and is strictly pro-inflammatory and can increase recruitment the pro-inflammatory interleukins IL-1, IL-6, and IL-8 (Victor & Gottlieb, 2002). Although largely produced by mononuclear phagocytes, TNF-α is also produced by T lymphocytes, Kupffer cells, neural cells, and endothelial cells can also produce TNF-α (Moldoveanu, Shephard, & Shek, 2001). In addition to TNF-α, the pro-inflammatory interleukins, which will be addressed later in this review, include IL-1 α and IL-1-β, IL-6.

Monocytes and macrophages are the primary sources of IL-1 α and IL-1-β but both are also released by endothelial cells, epithelial cells, and fibroblasts (Weber, Wasiliew, & Kracht, 2010). IL-1 α and IL-1-β are comprised of different amino acid sequences, but function similarly. IL-1 α is often bound to a membrane where it signals through the autocrine and juxtracrine pathways (Weber, Wasiliew, & Kracht, 2010). IL-1 β is secreted via unconventional methods by caspase-1, a cysteine protease and a regulator of inflammation (Keller et al., 2008). This pair of interleukins are involved in several scenarios including the onset of fever, the priming of T cells, and macrophage recruitment (Ben-Sasson et al., 2009; Copray et al., 2001; Sims & Smith, 2010). The IL-1 family has been used to monitor inflammation related to exercise training due to its response to exercise training. Circulating cytokines from the IL-1 family has been demonstrated to decrease after aerobic exercise (Goldhammer et al., 2005) and increase throughout a season when training volumes increase (Vaisberg, de Mello, Seelaender, dos Santos, & Rosa, 2007). Additionally, the IL-1 family has been shown to increase after resistance exercise and in some cases remain elevated for up to 5 days post stimulus (Dennis et al., 2004). IL-1 is thought to be strictly a pro-inflammatory cytokine and the first interleukin to be discovered, lending to the large body of research on the molecule (Hazuda, Lee, & Young, 1988). The characteristics of the IL-1 family lend to it being a popular option when quantifying inflammation. However, similar to CRP, the IL-1 family's half-life is not optimal for monitoring training, with the half-life of IL-1 α 15 hours and of IL-1 β 2.5 hours (Hazuda, Lee, & Young, 1988).

IL-6 has both pro- and anti-inflammatory characteristics. A cell surface must express the IL-6 receptor and the protein gp130 to allow IL-6 to bind and influence a cell (Rose-John, 2012). Gp130 is present on all cells while the membrane bound receptor for IL-6, IL-6 receptor (IL-6r), is only expressed on specific cells, primarily hepatocytes and leukocytes (Rose-John, 2012). IL-6 bound to its receptor combines with Gp130 to trigger a series of reactions termed the classic-signaling pathway leading to the anti-inflammatory mechanisms of IL-6 (Scheller et al., 2011). The classic singaling pathway involves the signaling the JAK/STAT, P13Km and ERK pathways (Rose-John, 2012; Scheller, Chalaris, Schmidt-Arras, & Rose-John, 2011). Stat3 activation leads to intestinal epithelial cell proliferation, inhibition of epithelial cell apoptosis and induction of hepatic acute phase response (Rose-John, 2012).

The pro-inflammatory actions of IL-6 are thought to be chronic in nature and result from trans-signaling (Scheller et al., 2011) and may increase the number of mononuclear cells in conditions such as rheumatoid arthritis and type II diabetes (Kaplanski, Marin, Montero-Julian, Mantovani, & Farnarier, 2003). Chronically elevated levels of IL-6 may play a role in decreasing insulin sensitivity as infusing IL-6 has been shown to impair the actions of insulin, while blocking IL-6 has increased insulin sensitivity in mice (Timper et al., 2017). A soluble form of IL-6R (sIL-6R) which can bind IL-6 to gp130 on a wider variety of cells, is thought to be responsible for the pro-inflammatory capabilities of IL-6 through the trans-signaling pathway (Kaplanski et al., 2003; Scheller et al., 2011). Trans-signaling will activate the immune system, recruiting mononuclear cells, decreasing T-cell apoptosis, and downregulating regulatory T-cell differentiation (Rose-John, 2012). IL-6 can activate endothelial cells, causing them to secrete IL-8, through the IL-6-sIL-6R pathway, which recruits and activates neutrophils locally (Kaplanski et al., 2003). Research focused on the response of IL-6 to exercise has demonstrated that it tends to increase after a single bout of exercise regardless of modality and seems to be positively associated with exercise volume and intensity (Moldoveanu, Shephard, & Shek, 2001). Some studies have found that IL-6 returns to baseline within 6 hours post exercise (Moldoveanu, Shephard, & Shek, 2001), but it may also remain elevated for 72 hours after exercise (Smith et al., 2000). Conflicting responses between IL-1 and IL-6 have also been shown, where IL-1 has remained at baseline or significantly decreased while IL-6 significantly increased after exercise (Smith et al., 2000; Ullum et al., 1994). The kinetics of IL-6 present a similar issue as IL-1 when monitoring multiple training sessions a week, with the possibility of remaining significantly above baseline throughout multiple sessions (Smith et al., 2000). The pro- and anti-inflammatory roles of IL-6 may make it difficult to distinguish which mechanisms are in action when levels are elevated.

TNF-α is a pro-inflammatory cytokine involved in the acute phase response. It has been found to increase inflammation by activating IL-1, IL-6, IL-8, and NF-κB (Victor, & Gottlieb, 2002). NF-κB helps regulate the immune system by promoting the transcription of pro-inflammatory cytokines, adhesion molecules, and iNOS (Moynagh, 2005). The strictly pro-inflammatory nature of TNF- α provides a less convoluted understanding of the physiological processes resulting from elevated levels, when comparing it to IL-6. TNF-α tends to remain elevated above baseline hours after an exercise session (Ostapiuk-Karolczuk et al., 2012; Reihmane, Jurka, Tretjakovs, & Dela, 2013) and may remain elevated greater than 28hrs post exercise (Ostapiuk-Karolczuk et al., 2012). There is also evidence demonstrating no significant changes to baseline after exercise (Chen et al., 2009).

The cytokines described above have been used to quantify the immune systems inflammatory response to a single exercise session. While CRP is widely used to assess inflammation in response to exercise, these cytokines may provide advantages when compared to CRP as they tend to peak and return to baseline more rapidly. For instance, IL-6 has been shown to increase above baseline immediately after exercise, while CRP remains at baseline (Costello et al.,2018). Similarly, it has been demonstrated that IL-6 increases above baseline at 30 minutes, 120 minutes, and 24 hours post exercise while CRP remains unchanged (Gordon III et al., 2017). However, TNF-α, IL-β, IL-6, and IL-10 have been shown to increase significantly after a marathon (Moldoveanu, Shephard, & Shek, 2001), suggesting that both pro- and anti-inflammatory pathways are stimulated simultaneously, presenting a problem since IL-6 can suppress the synthesis of TNF-α and IL-1 by macrophages, through IL-6R (Moldoveanu, Shephard, & Shek, 2001). This may explain some conflicting results in studies. A study by Donges, Duffield & Drinkwater (2010) recruited with sedentary subjects and found a decrease in baseline CRP for resistance and aerobic training groups over a 10-week period, but no change in IL-6 in either group. A second study conducted by Mitchell et al. (2013) utilized a 16-week resistance training protocol, finding that baseline CRP was elevated after the procedure, with no differences in resting IL-6 or TNF-α. This may have been a result of the researchers not allowing enough time for CRP to return to baseline (96hr), before making the final blood draw. The contradictory findings of these markers make it difficult for sport scientists to quantify inflammation in athletes, resulting in the requirement for a more consistent monitoring tool.

Creatine Kinase

CK is an enzyme found in skeletal and cardiac muscle, photoreceptors, and neurons (Wallimann et al., 1998). CK assists in the regeneration of ATP by catalyzing the transfer of the N-phosphoryl group from ADP (Wallimann et al., 1998). The phosphorylation of ADP to ATP allows the cell to maintain the necessary energy homeostasis (Wallimann et al., 1998). CK can leak out of a cell and into the blood stream. Changes in serum CK can result from tissue damage and cellular necrosis in response to acute and/or chronic muscle injury (Adams, Schechtman, Landt, Ladenson, & Jaffe, 1994). CK has been used to detect myocardial infarctions, leaking into the blood from the cardiac muscle when damage occurs (Adams et al., 1994). Additionally, serum CK also increase due to exercise (Brancaccio et al., 2007). Elevated serum CK levels from skeletal muscle has been shown to be related to delayed onset muscle soreness (DOMS) (Doma et al., 2018; Raeder et al., 2016). The connection between CK and DOMS may be related to the damage done to z-discs during exercise training. Z-discs anchor the actin filaments to the sarcomere and must support high forces when the actin is coupled to myosin (Bellin, Huiatt, Critchley, & Robson, 2001). Muscle damage often occurs at the z-discs due to the forces they must sustain (Nguyen et al., 2009). The pain and immobility associated with DOMS is positively related to the severity of z-disc disruption (Nosaka, Lavender, Newton, & Sacco, 2003). Z-disc tears also compromise the ability of muscle to produce force. Without the full integrity of their anchors, actin is unable to produce maximal force when coupled to myosin (Nguyen et al., 2009). Training results in increased integrity of the z-discs, allowing them to endure higher exercise volume with less breakdown. (Hody et al., 2011).

Elevated serum CK is related to the volume of eccentric exercise completed (Hody et al., 2011). Muscles are capable of withstanding higher forces during the eccentric phase of a movement, but the shearing nature of myosin from actin during eccentric contractions causes muscle damage (Cheung, Hume, & Maxwell, 2003; Nosaka et al., 2003). CK has been shown to increase with greater exercise intensity, volume, and duration, although, only a moderate body of evidence exists that supports this (Horta et al., 2017). There may be stronger evidence supporting subjective measures of acute training load (TL), such as the DALDA and POMS, as accurate assessments of fatigue when compared to CK. Some studies show subjective measures hold a stronger relationship to chronic TL, while CK may not display any changes over time (Saw, Main, & Gastin, 2015). Studies have also shown an increased TL, measured by session RPE (sRPE), increased CK with no corresponding decrease in counter movement jump height (Coutts, Reaburn, Piva, & Rowsell, 2007; Freitas, Nakamura, Miloski, Samulski, & Bara-Filho, 2014). Byrne and Eston (2002) found a decrease in countermovement jump, squat jump, and drop jump with significantly elevated serum CK level three days after recreationally trained subjects performed 10 sets of 10 repetitions of the back squat at 70% of body weight. However, Twist and Eston (2005) found peak power to be significantly decreased 72 hours after an exhaustive training session on a cycle ergometer, with CK only remaining elevated for 48 hours after the protocol. These examples demonstrate that the literature is divided on utilizing CK to predict the readiness to performance.

Similar to CRP, CK may remain elevated 48hrs and longer after exercise (Ascensão et al., 2008; Byrne & Eston, 2002; Twist & Eston, 2005). The delay in response time makes utilizing serum CK levels as a monitoring tool in athletes impractical. There is also evidence showing decreases in CKs response as an athletes training level increases, decreasing the sensitivity of the marker to changes in training load in well-trained individuals (Clarkson, Nosaka, & Braun, 1992; Saw et al., 2016).

Cell Free DNA

Historically, cf-DNA has been used primarily in medicine and very little in exercise physiology related scenarios, although the body of exercise related literature is increasing due to rapid response time. cf-DNA has multiple sources including, necrosis, apoptosis, and the release of NETs (Elshimali et al., 2013). The primary source of cf-DNA from exercise is hypothesized to result from the production of reactive oxygen species during high intensity and prolonged exercise (Hopker et al., 2016) which may trigger the release of NETs, explaining the fast time-course response of cf-DNA (Beiter, Fragasso, Hudemann, Nieß, & Simon, 2011).

Investigations of cf-DNAs' response to exercise is limited, resulting in the need to examine cf-DNA related medical findings to understand the markers kinetics. The hydrolysis of cf-DNA occurs by DNAse 1 nuclease, resulting in a 157.6-minute half-life (Wilson et al., 2017). The short half-life ofcf-DNA results in a rapid acute response when compared to other biochemical markers discussed in this review. Wilson et al. (2017) observed that cf-DNA returned to baseline within 6 hours in patients who underwent surgery, while research using exercise as a stimulus presented a return to baseline of 30 minutes post (Atamaniuk et al., 2010; Beiter et al., 2011). There are multiple sources of cf-DNA, of which the Alu repeat family is the most common (Elshimali et al., 2013). Alu sequences are considered short interspersed elements (SINEs), containing about 300 base pairs (Elshimali et al., 2013). The number of base pairs are thought to have a relationship to the source. The common Alu is approximately 180 to 1,000 base pairs in length and may appear in plasma as a result of apoptosis, while necrosis produces pairs of 10,000 (Elshimali et al., 2013).

Baseline plasma levels of cf-DNA can differ between individuals, but group mean concentration have been reported. Yoon et al. (2009) used quantitative PCR (qPCR) to measure cf-DNA and found a baseline of 10.4 (range, 1.6 to 89.8 ng/ml) in their control group of 105 healthy subjects. Other studies that include baseline cf-DNA concentrations of control groups have found similar levels, suggesting that 10.4 ng/ml is an accurate representation (Jahr et al., 2001; Rhodes, Wort, Thomas, Collinson, & Bennett, 2006), while baseline serum levels of cf-DNA may be higher due to the release of DNA during clotting (Holdenrieder et al., 2005). It is important to note the measurement method when comparing results from different studies, as measurement sensitivity can differ between instruments (Breitbach et al., 2012). The target gene used for qPCR can also influence measured concentrations, particularly when comparing the common short sequences to longer sequences (Breitbach et al., 2012).

The medical field often uses cf-DNA in acute and chronic cases, to either quantify the level of damage, or to monitor a condition over time. Apoptosis and necrosis have both been shown to increase in myocardial infarction patients (Chang et al., 2003). A corresponding increase in cf-DNA can be expected, as well as increases in CK levels, although the time-course response of the two markers is different (Chang et al., 2003). Elevated levels in acute conditions such as myocardial infarction suggests that cf-DNA responds quickly to cell damage. This is further supported by the response of cf-DNA resulting from blunt trauma. Lo et al. (2000) collected blood samples from patients with a median of 60 minutes post trauma showed significantly elevated serum cf-DNA concentrations when compared to controls.

Assessing changes to baseline cf-DNA levels is a useful tool for cancer detection and monitoring (Spindler et al., 2016). One hundred and seventy-five breast cancer patients displayed a mean cf-DNA value of 105.2 ng/ml measured via qPCR, which was significantly higher than the control group with a mean of 77.06 ng/ml (Catarino et al., 2008). Using cf-DNA to determine the stage and prognosis of cancer progressed due to the ability of researchers to identify the genotype of certain cancers, only targeting the cf-DNA sequences of importance in a case by case basis (Diaz Jr & Bardelli, 2014). Researchers can now monitor tumor characteristics in real time (Diaz Jr & Bardelli, 2014). It is also possible to locate a tumor and track its response to therapeutic modalities using its genomic signature. Currently, specific cf-DNA sequences have been identified for cancers of the central nervous system, breast, cervical, ovarian, esophageal, stomach, and more (Elshimali et al., 2013). The ability to identify circulating DNA sequences from specific sources is expected to continue to improve (Elshimali et al., 2013). This has implications beyond the study of cancer, possibly identifying cf-DNA from specific regions and tissues in the body, allowing a more accurate quantification of local damage.

Acute cf-DNA increases have been shown to occur immediately after exhaustive treadmill running. Compared to baselines, levels of cf-DNA increased 9.9-fold immediately after the run and fell to 4.1-fold higher after 30 minutes (Beiter et al., 2011). This demonstrates the short half-life of cf-DNA resulting from exercise, strengthening the case for its utility in an athlete monitoring program. The resulting increase was shown to be from nuclear DNA rather than mitochondrial DNA (cf-mtDNA) (Beiter et al., 2011). Exercise training may cause adaptations that blunt cf-mtDNA response over time. Ten weeks of moderate intensity aerobic exercise has been shown to increase baseline cf-mtDNA concentrations, which was correlated with ($r = 0.6$) with an increase in VO_2 max ($r = 0.6$) (Wang, Hiatt, Barstow, & Brass, 1999). The increased VO_2 max may be due to an increase in mitochondria (Hood, Chabi, Menzies, O'Leary, & Walkinshaw, 2007; Little, Safdar, Wilkin, Tarnopolsky, & Gibala, 2010), resulting in more mitochondrial cell turn-over (Lim et al., 2000). Correlations between VO_2 max and baseline cf-mtDNA has been demonstrated in a previous study as well, suggesting higher baseline cf-mtDNA levels may be indicative of higher aerobic fitness (Wang et al., 1999).

Few studies have been conducted on the effect resistance exercise on cf-DNA concentration. Of the investigations to explore resistance exercise and cf-DNA, two studies focused on levels immediately after training, while two other investigations measured cf-DNA 48 hours or more after a session. A study conducted by Atamaniuk et al. (2010) recruited national class weightlifters and found that cf-DNA significantly increased immediately after 6 sets of 1 to 5 repetitions with 100% relative intensity, using 6 multi-joint lifting exercises, and returned to baseline within two hours. A significant increase in serum leukocytes, neutrophils, monocytes, and thrombocytes were shown immediately after and two hours post (Atamaniuk et al., 2010). Similar results were shown when trained men completed 3 days a week of 2 sets of 8 multi-joint resistance exercises for 4 weeks. Subjects were separated into three groups, two groups completed 2 sets of 5 repetitions at 60% of their 1RM, while the remaining group completed as many reps as possible at 90% of their 1RM (Tug et al., 2017). Researchers drew blood samples from the finger after each set, and before and after each training session. The long interspersed nucleic element (LINE) L1PA2 family was used as the target gene (Tug et al., 2017), which is roughly 6,000 base pairs in length (Sheen et al., 2000). All groups showed a significant elevation in cf-DNA on all training days after set 12 through 16, when compared to baseline (Tug et al., 2017). Although the high-intensity group had the greatest increases, the results were not significantly different compared to the other groups. There were also significant differences in cf-DNA measured prior to exercise on training days 2, 5, 8, and 12 when compared to baseline (all 48 hours post), with little variation between groups (Tug et al., 2017). The lack of changes in cf-DNA between groups prior to a training session may correspond to the similarity of each week in the study. Adding training variations to one group may have resulted in fluctuations between the samples taken prior to the following session. These studies suggest cf-DNA is correlated with exercise intensity. A future study examining the capability of cf-DNA to predict acute performance changes, due to high intensity resistance training, would be useful in athletic monitoring.

Measuring the chronic response of cf-DNA may prove to be more difficult. There is not much research, in any discipline, that is able to provide insight on how it reacts to a stimulus presented 48 hours prior and longer. Fatouros et al. (2006) conducted a 12-week resistance training protocol to examine the responses of cf-DNA, CK, CRP, and uric acid (UA). Researchers split the time into four training periods: t1 and t4 were low-volume, t2 was high-volume, and t3 was very-high-volume. Subjects were asked to refrain from resistance exercise for a period of 8 weeks prior to the study, to achieve a baseline level of cf-DNA (Fatouros et al., 2006). Researchers drew blood 96 hours after the conclusion of each training period. They found f-DNA levels to be significantly elevated compared to baseline at t1, t2, and t3 (highest), and had returned to baseline at t4. Cf-DNA was correlated with training volume at t2 ($r=0.755$) and t3 (r=0.786), no other correlations between training volume and serum markers were found (Fatouros et al., 2006). CK was significantly elevated at t3 only, CRP at t2 (300%) and t3 (400%), and UA at t2 and t3 (Fatouros et al., 2006). Performance measures increased slightly from baseline to t1, t1 to t2 (highest), decreased from t2 to t3 (lowest), and increased from t3 to t4 (Fatouros et al., 2006). The increase to baseline performance measures could be explained by the 8-week break prior to the study. The 3-week low volume training period may have increased motor unit recruitment, further increasing performance variables (Deschenes et al., 2000). T3 represents the very-high-volume period, where a decrease in performance due to fatigue would be expected (Izquierdo et al., 2009). Cf-DNA, CK and CRP were all highest at this point, and reflected the performance decrease (Fatouros et al., 2006). The source of the cf-DNA found in this study may be from necrosis, since samples weren't taken until 96hrs after exercise. It would have been informative to have longer sequences, as they are related to necrosis, along with the 189 base pair sequence used in this study (Breitbach et al., 2012).

Gentles et al. (2017) completed a similar study using weightlifters. Researchers tracked volume-load-displacement (VLD) and training-intensity displacement (TID) over a period of 20 weeks. The study was divided into 6 training phases, and blood was drawn to assess cf-DNA, CRP, and CK 48 hours after each phase. Significant changes occurred in VLD and TID, throughout the study (Gentles et al., 2017). CK %Δ was correlated to the VLD mean for the most recent 4 weeks (4wk) (r=0.86) and VLD4wk %Δ (r=0.86), while TID1wk was correlated to CRP (r=0.83) (Gentles et al., 2017). Cf-DNA %Δ was correlated with CRP and CRP %Δ (r = 0.87, r = 0.86) (Gentles et al., 2017). The authors stated that the correlation between CK %Δ and VLD4wk may be misleading, since CK was not related to the VLD of the most recent week. The results showed that CK %Δ did not correspond to VLD Δ or TID Δ, while cf-DNA and CRP appeared to rise and fall simultaneously with VLD and TID (Gentles et al., 2017). The weightlifters may have been accustomed to the movements they were completing, which could explain the different results between chronic changes in cf-DNA seen in the study by Gentles et al. and Fatouros et al. between the two studies. Less muscle damage occurs when individuals have had time to adapt to the new stimulus (Skurvydas, Brazaitis, & Kamandulis, 2011). Adding measures to track performance variations as VLD or TID increase, while cf-DNA and CRP are not significantly affected, would provide insight on whether cf-DNA can be used to predict a trained athletes readiness to compete.

Conclusion

Skeletal muscle damage and inflammation are important mediators in the response and adaptation to resistance training. Periods of high training volume and/or intensity lead to elevated damage and inflammation, which produces greater improvements to muscle strength, and especially hypertrophy, compared to low training volumes and/or intensities (Bloomer et al., 2010; Gormley et al., 2008; Kramer et al., 1997; Wilborn et al., 2009). Quantifying muscle damage and inflammation can provide information regarding the current fatigue state of an individual, including whether their current fatigue state is elevating their risk for disease (Booth et al., 2006; Brancaccio et al., 2007). CRP, IL-1, IL-6, TNF-α, and CK are biochemical markers with numerous publications exploring their responses to resistance training and usefulness in athlete monitoring (Souglis et al., 2015). The kinetics and inconsistencies found within and between these popular markers leave a gap in tracking fatigue biochemically that has yet to be filled (Chen et al., 2009; Mitchell et al., 2013; Saw et al., 2015; Smith et al., 2000; Ullum et al.,1994). cf-DNA shows promise as a biochemical marker to monitor fatigue due to its short half-life and sensitivity to changes in resistance training volume (Breitbach et al., 2012; Tug et al., 2017). Further research is required to explore the relationship between cf-DNA and different training modalities, and to gain a better understanding of the kinetics of cf-DNA in response to varying resistance training volume-loads.

CHAPTER 3. THE RELATIONSHIP BETWEEN CELL FREE DNA, CREATINE KINASE, C-REACTIVE PROTIEN, AND DELAYED ONSET MUSCLE SORENSS

INTRODUCTION

Biochemical markers play an important role in exercise physiology research and athlete monitoring. Two of the most commonly used markers, Creatine Kinase (CK) and C-reactive protein (CRP), provide researchers and sport scientists with an understanding of the immune response subjects and athletes have to a specific training stimulus (Booth, Probert, Forbes-Ewan, & Coad, 2006; Freitas, Nakamura, Miloski, Samulski, & Bara-Filho, 2014; Souglis et al., 2015). Understanding an individuals' immune response to a given training session can provide information regarding resulting fatigue and performance decrements from the training bout. (Freitas et al., 2014). Increases in baseline CK, CRP, and delayed onset muscle soreness may occur after resistance training and are positively associated with higher volume loads(Horta, Bara Filho, Coimbra, Miranda, & Werneck, 2019; Calle & Fernandez, 2010; Shanely et al., 2014). Performance variables, such as vertical jump height, tend to be negativily associated with increases to resistance training volume load (Doma et al., 2018). The use of quantifying the acute phase responses of CK and CRP following resistance training has been outlined in the literature. While these biochemical markers have proven useful for tracking muscle damage and immune response from training and athlete monitoring, they present limitations that lower the practicality of their use in applied conditions.

CK and CRP are restricted by half-lives lasting about 30 hours and 19 hours respectively (Pepys & Hirschfield, 2003; Rogers, Stull, & Apple, 1985). The described half-lives can result in the marker remaining elevated 72 hours or more after a resistance training bout for CK and 48 hours or longer in the case of CRP (Barquilha et al., 2011; Rodrigues et al., 2010). A marker remaining elevated for 72 hours or longer is not helpful when quantifying the immune response to a single exercise stimulus, given that it is unlikely an athlete will only complete a single training bout within that period (Gentles et al., 2017). The extended elevation of CK and CRP requires the cessation of training to ensure marker concentrations reflect the session of interest, rather than a result of additional sessions or accumulated fatigue. The lack of practicality with use of CK and CRP has led researchers and sport scientists to explore other biochemical markers as options for monitoring athletes' immune responses.

Cell Free DNA (cf-DNA) is a biochemical marker used for quantifying the immune response and function of patients with varies chronic conditions or who have experienced trauma or surgery (Catarino et al., 2008; Chang et al., 2003; Wilson et al., 2017). More recently, cf-DNA has demonstrated promise in the field of exercise physiology and for athlete monitoring. The short half-life of cf-DNA, 157.6-minute half-life in clinical studies (Wilson et al., 2017), and as short as 30 minutes after an exercise stimulus (Atamaniuk et al., 2010; Beiter, Fragasso, Hudemann, Nieß, & Simon, 2011), results in a fast return to baseline, within two hours in some studies using exercise (Atamaniuk et al., 2010; Atamaniuk et al., 2004).The favorable kinetics of cf-DNA for athlete monitoring may be attributable to neutrophil extracellular traps (NETs), which have been proposed to be primary source of cf-DNA following exercise training (Breitbach, Tug, & Simon, 2012; Haller et al., 2018). Neutrophils rapidly release NETS following a stimulus tissue damage to aid in the clearing of pathogens and damaged tissue (Kolaczkowska & Kubes, 2013; Margraf et al., 2008).

Currently, there is no research describing the relationship between cf-DNA, CK, CRP, performance decrements, and perceived DOMS. The purpose of this investigation was to examine the relationship between cf-DNA, creatine kinase (CK), C-reactive protein (CRP), vertical jump testing, and delayed onset muscle soreness (DOMS) after a high-volume resistance training bout.

METHODS

Experimental Approach to the Problem

Seventeen participants were recruited for this study. Participants reported to the lab for familiarization of the study. Self-reported 1 repetition maximums (1RM) for the back squat and bench press exercises were collected from the participants, and body mass measured using a digital scale (Tanita B.F. 350, Tanita Corp. of America, Inc., Arlington Heights, IL). A previous damage inducing resistance training protocol was slightly modified and used for this study (McBride et al., 2009).

The study design is outlined in Table 1. Participants were divided into two groups, one group that received branched chain amino acids (BCAAs) and one control group that received a placebo. The supplement group ingested 0.17g of the amino acid supplement for every kilogram of bodyweight prior to the training session (Ajipure Fusi-BCAA pharmaceutical grade, made by Ajinomoto) with 20oz of water mixed with a non-caloric sugar-free flavor (Propel zero calorie mix). The supplement contained 2.5g of Leucine, 1.25g L Isoleucine, and 1.25g L-Valine per 5 grams. The control group consumed the 20oz of water with the sugar free flavoring. Each group consumed their respective drinks 24 hours prior and immediately prior to the training session, and again 24 hours post and 48 hours post. Participants consumed an additional dosage during their exercise session on the training day. Blood was drawn to assess serum cf-DNA, CK, and CRP levels prior to the training session, with cf-DNA collected again immediately post, and CK and CRP at 24- and 48-hours post. Point specific muscle soreness on a scale of 1 to 10 was collected on day 2 prior to training, day 3, and day 4. Static jump height (SJH), countermovement jump height (CMJH), and the Bosco repeated jump test (BOSCO) were collected on day 1, day 3, and day 4.

Table 3.1

Outline of Study Design

Familiarization	Day 1	Day 2	Day 3	Day 4
• Self-reported 1RM • Body Weight	• Hydration • BCAAs/Placebo • Warm-up • Jump Testing • BOSCO	• Hydration • Muscle soreness chart • BCAAs/Placebo • Pre-session blood draw • Warm-up • Training and BCAAs/Placebo • Post session blood draw	• Hydration • 24hr post blood draw • Muscle soreness chart • BCAAs/Placebo • Warm-up • Jump testing • BOSCO	• Hydration • 48hr post blood draw • Muscle soreness chart • BCAAs/Placebo • Warm-up • Jump Testing • BOSCO

Participants

Seventeen college age males (18-32 years old) were recruited for this study. All participants had a minimum of one year of resistance training. Training loads were quantified using the participants self-reported 1RM for the back squat (148.4 kg ± 39.6 kg) and bench press (104.3 kg ± 22.2 kg). Participants were randomly assigned to either the BCAA (n = 8; age = 26.2 ± 3.5yrs; body mass = 90.9 ± 8.5) or control group (n = 9: age = 24.0 ± 3.3yrs; body mass = 95.3 ± 13.7kg). The NSCA Health/ Medical Questionnaire was used to assess participant health status. No significant differences in strength or anthropometric measures were found between groups, and no health concerns were reported that would make participants ineligible. All participants were asked to refrain from physical exercise for a period of three days prior to day 1 of the study. Participants were notified of the potential risks and benefits of the study and were provided and signed an informed consent document. This study was approved by the East Tennessee State University Institutional Review Board.

Training Protocol

Participants completed a pre-testing warm-up consisting of 25 jumping jacks and 3 sets of 5 repetitions of mid-thigh pulls at 40 kg and at 60 kg. Each participant performed 5 sets of 10 repetitions of both the back squat and bench press at 57% of their estimated 1RM. Participants were allowed 2 to 3 minutes rest between sets and 3 min of rest after the back squat before beginning the bench press.

Biochemistry

Venous blood draws were performed pre, immediately post, 24 hrs post, and 48 hrs post resistance training. Blood was drawn from the median antecubital vein into 9 mL vacutainers (BD serum separator). cf-DNA was assessed pre and immediately post resistance training, while CK and CRP were measured pre, 24- and 48-hours post resistance training. Blood was centrifuged at 6,000 RPMs for 15 minutes (SmithKline Beecham VanGuard 6000 centrifuge, Haddonfield, New Jersey) and stored in a -86°C (Thermo Fisher Scientific, Waltham, MA). Serum was thawed and CK and CRP was measured from serum in duplicate using an automated immunoassay analyser (IMMULITE 1000, Siemens Healthcare, Erlangen, Germany). cf-DNA was isolated from serum using QIAamp MinElute ccfDNA Midi Kits (Qiagen, Germany) (Ok et al., 2020). cf-DNA was then quantified via qPCR (Bio-Rad CFX96 Touch™ Real-Time PCR detection system, München, Germany) (Tug et al., 2017) with primers copying the RNase-P gene (Malik, Shahni, Rodriguez-de-Ledesma, Laftah, & Cunningham, 2011; Marcuello et al., 2005).

Performance Variables

SJH and CMJH were collected after a generalized warm-up. Each jump was tested using two conditions, a PVC pipe and a 20kg barbell. After two warm-up jump trials, participants were instructed to stand on the force plates and given the command "three… two …one… jump", at which point they attempted to reach maximum jump height. Participants were allowed two trials of each jump under both conditions; one minute of rest was provided between trials. The average of the two jumps was reported. If the two trials were 2cm apart in height or more, a third trial was attempted and the average of the closest two jump heights were used for analysis.

The BOSCO consists of repeated jumps for 30 seconds. All jumps began from a squat depth with a ~90° knee angle. Participants were encouraged to achieve maximum height with each jump. The test-retest reliability for SJ, CMJ, and BOSCO tests yield ICCs of 0.99, 0.88, and 0.87, respectively (Acero et al. 2011; Dal Pupo et al. 2014; Slinde, Suber, Suber, Edwén, and Svantesson, 2008). Dual force plates were used for all data collection, with a sampling rate of 1,000 Hz (Rough Deck HP; Rice Lake, WI), implementing the impulse method to measure jump height (Street, McMillan, Board, Rasmussen, & Heneghan, 2001). LabVIEW software was used for all jump data analysis (LabVIEW 2017, National Instruments).

Assessment of Soreness

Participants returned to the lab to describe their perceived muscle soreness 24 and 48hrs after the session. A chart was provided to participants, which divides the body into sections, and each participant was asked to provide a level of soreness on a scale of one to ten (Figure 1) (Sands, McNeal, Murray, and Stone, 2015). The body sections included on the soreness chart was: anterior legs (quads, posterior legs (hamstrings), glutes, lower back (erector spinae, quadratus lumborum), upper back (trapezius, rhomboids), shoulders (deltoid groups), chest (pectoralis major and minor), and abdominal (rectus, transverse, obliques). DOMs was assessed by summing the responses of all point specific soreness ratings.

Figure 3.1

Point specific muscle soreness chart

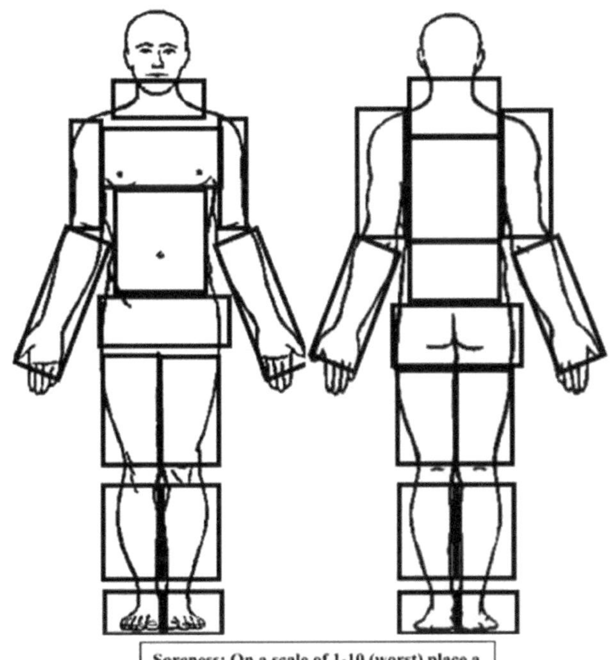

Statistical Analysis

Data were analyzed with SPSS (Version 25.0. Armonk, NY: IBM Corp). Descriptive

statistics in the form of means and standard deviations were calculated for all variables. Pearson

correlations were used to assess the relationship between relative changes (percent change) in cf-

DNA (%Δ cf-DNA) pre to immediately post to percent changes (%Δ) in CRP, CK, SJH, CMJH,

and BOSCO pre to 24hrs and pre to 48hrs. The relationship between DOMS pre to 24hrs and pre

to 48hrs and %Δ cf-DNA pre to immediately post, CK pre to 24hrs and pre to 48hrs, and CRP

pre to 24hrs and pre to 48hrs were assessed via Spearman's rho. All correlations were completed

for the non-supplement group, supplement group and both groups combined. The critical alpha level for all analyses was set at $p \leq 0.05$. All correlation coefficients were interpreted as 0.00 to 0.09 = negligible, 0.10 to 0.39 = weak, 0.40 to 0.69 = moderate, 0.70 to 0.89 = strong, 0.90 to 1.00 = very strong (Schober, Boer, & Schwarte, 2018)

RESULTS

The correlation between %Δ DOMS 48hr and %Δ CRP 48hr for the non-supplement group was the only model found to be significant ($r = 0.71$; $p = 0.02$). Table 3.2 contains descriptive statistics in the form of mean ± SD for all variables for the supplement, non-supplement and combined groups.

Table 3.2

Deceptive Statistics in the Form of Means ± SD for All Variables

	Combined	Supplement	Non-Supplement
%Change cf-DNA	72.91 ± 71.27	56.11 ± 60.56	91.81 ± 77.42
%Change 24hr CK	5.38 ± 60.73	13.86 ± 65.56	-4.16 ± 53.20
%Change 48hr CK	-6.34 ± 35.36	4.58 ± 35.42	-18.62 ± 30.99
%Change 24hr CRP	14.08 ± 49.24	10.46 ± 19.18	18.14 ± 68.61
%Change 48hr CRP	5.34 ± 36.89	18.41 ± 31.00	-9.37 ± 37.44
%Change 24hr Soreness	33.97 ± 39.53	29.61 ± 36.94	38.87 ± 41.71
%Change 48hr Soreness	20.64 ± 41.88	15.63 ± 41.25	26.27 ± 41.88
%Change 24hr BOSCO	2.80 ± 4.23	0.42 ± 2.49	5.48 ± 4.18
%Change 48hr BOSCO	2.92 ± 6.14	1.31 ± 6.07	4.73 ± 5.68
%Change 24hr CMJ 0kg	3.26 ± 8.63	0.34 ± 6.89	6.56 ± 9.19
%Change 48hr CMJ 0kg	1.58 ± 9.45	-1.98 ± 10.42	5.58 ±6.10
%Change 24hr CMJ 20kg	8.50 ± 14.58	2.75 ± 10.52	14.96 ± 15.77
%Change 48hr CMJ 20kg	5.89 ± 12.18	2.56 ± 10.15	9.64 ± 13.14
%Change 24hr SJ 0kg	3.94 ± 11.98	2.41 ± 13.54	5.66 ± 9.66
%Change 48hr SJ 0kg	0.57 ± 9.22	1.26 ± 11.04	-0.20 ± 6.52
%Change 24hr SJ 20kg	8.29 ± 17.18	3.47 ± 10.52	13.71 ± 21.15
%Change 48hr SJ 20kg	3.48 ± 9.43	0.04 ± 8.30	7.34 ± 9.11

DISCUSSION

The purpose of this investigation was to examine the relationship between cf-DNA, creatine kinase, C-reactive protein, vertical jump testing, and delayed onset muscle soreness. The purpose was accomplished by implementing a resistance training protocol with the goal to induce muscle damage and inflammation in participants, while tracking the relationship between changes to biochemical, performance, and DOMS. The current study is the first to explore the relationship between cf-DNA, performance variables, subjective survey data, and other biochemical markers (CK, CRP).

The lack of significant findings in this study may be partly attributable to the limitation of a small sample size. The lack of association between these variables may in part be due to the small sample size, as correlations tend to stabilize with a greater number of participants (Schönbrodt & Perugini, 2013). The large standard deviations for %Δ DOMS 24hr and %Δ cf-DNA, found to be $33.87 \pm 39.53\%$ and $72.91 \pm 71.27\%$, respectively further illustrates the need for a larger sample size. Unfortunately, the problem of small sample size is common in the sport science literature (Gentles et al., 2017).

The combined group means at 24 and 48hrs for BOSCO, CMJ 0 and 20kg, and SJ 0 and 20kg increased from baseline by margins ranging from 0.57% to 8.50%. The relative increase in performance variables was unexpected, as research tends to show a decrease in jump height and fatigue index measured by the BOSCO repeated jump test in response to a high resistance training stimulus (Sams, Sato, DeWeese, Sayers, & Stone, 2018; Sever et al., 2017). Most of the participants in this study were unfamiliar with any form of jump testing, which may have resulted in a familiarization effect following baseline measurements. It is also possible that the volume-load did not result in enough fatigue to negatively influence performance variables.

The relative changes to CK and CRP were lower than expected for a resistance training protocol deigned to induce skeletal muscle damage. CK was found to increase by 5.38% at 24hrs and was 6.34% lower from baseline at 48hrs. CRP was elevated by 14.08% at 24hrs and 5.34% at 48hrs. Da Silva et al. (2009) used a similar resistance training protocol designed to induce strength and muscle adaptations, consisting of 3 sets of 10 repetitions of the back squat, at 50, 75, and 100% of participants measure 10RM. Researchers found a ~50% increase in CK 48hrs after the protocol compared to baseline values. A second study implemented a 4x10 eccentric training protocol including 3 exercises and found an average increase of 1907.17% in 6 of their 8 participants (Neme Ide, Alessandro Soares Nunes, Brenzikofer, & Macedo, 2013). The discrepancies between the results from this study compared to the two studies mentioned could be due to multiple variables. The mean relative back squat strength of the participants in this study was 1.56, while the cohort recruited by Da Silva et al. (2009) had a mean relative strength of 1.27 for the back squat. Lower relative strength is associated with lower training age and an increased sensitivity of CK and CRP to resistance training (Neme Ide et al., 2013; Till, Darrall-Jones, Weakley, Roe, & Jones, 2017; Vincent & Vincent, 1997). A greater response of Ck and CRP can be expected through an unaccustomed exercise stimulus compared to exercise modalities that are commonplace in a participants training (Kasapis, & Thompson, 2005; Neme Ide et al., 2013; Vincent & Vincent, 1997). Da Silva et al. (2009) required that participants rest for 7 days before beginning their protocol, while the current study only mandated a 3-day rest period. Allotting participants a rest period over twice the duration may have resulted in a more novel back squat stimulus in the study conducted by Da Silva et al. A final explanation for the lack of variation found in CK and CRP may be related to intensity and/or the use of reported 1RM values instead of directly measuring 10RMs. It is possible that participants inaccurately reported their 1RMs and using an equation to estimate a 10RM from a reported 1RM can result in additional error (Niewiadomski et al., 2008). A lack of accuracy in the prescribed weight in this study may have led to a lower than intended intensity, resulting in lower muscle damage and immune response.

It is important to note the baseline levels of cf-DNA appeared higher in this study compared to others in the literature. The greater baseline values may be due to isolating cf-DNA from serum rather than plasma. Greater values can be expected in serum compared to plasma due to release of additional cf-DNA in serum during the clotting process (Lui et al., 2002). The variations of cf-DNA in serum and plasma highlights the importance of implementing a consisted protocol when using the biomarker for athlete monitoring.

CONCLUSION

Currently, the kinetics and consistency of the commonly used biochemical markers make them problematic for the purpose of athlete monitoring. The response of CK in this study failed to generate meaningful information. CRP was correlated with soreness for the non-supplement group at 48hrs, but only at the 48hr mark ($r = 0.71$; $p = 0.02$), which is impractical for athlete monitoring. While research has demonstrated inconsistencies in the acute phase response of CK and CRP, other explanations for the lack of significant correlations may be related to small sample size, familiarity of the training stimulus, and the nature in which intensity was prescribed. Self-reported DOMS and cf-DNA were the only two variables that varied from baseline as expected after implementing a high-volume resistance training protocol. Further research is required to determine if DOMS and cf-DNA are related in different participant cohorts with larger sample sizes.

CHAPTER 4. THE RELATIONSHIP BETWEEN CELL FREE DNA AND VOLUME LOAD

<u>INTRODUCTION</u>

Circulating cell free DNA (cf-DNA) was first outlined in literature by Mandel and Metais

in 1948 (Tsang & Dennis Lo, 2007). The use of cf-DNA has since been implemented to track the

immune response of numerous conditions, including predicting post-trauma organ failure (Lo,

Rainer, Chan, Hjelm, & Cocks, 2000), quantifying cell necrosis following myocardial infarction

(Chang et al., 2003), and monitoring tumor metastasis in cancer patients (Schwarzenbach et al., 2009). More recently, cf-DNA has been used in the field of exercise physiology to monitor the immune state of research participants and athletes (Gentles et al., 2017).

Resistance training may lead to an immune response due to local damage and/or ischemia resulting from the session (Atamaniuk et al., 2004; Gordon III et al., 2017), resulting in elevated cf-DNA (Breitbach, Tug, & Simon, 2012). The inflammatory response to damage and low blood delivery leads to innate immune system response, including increased neutrophil recruitment to an area (Freidenreich & Volek, 2012).

Neutrophil extracellular traps (NETs), rather than cell necrosis or apoptosis, are thought to be the primary source of cf-DNA following exercise training (Breitbach, Tug, & Simon, 2012; Haller et al., 2018). NETs are released rapidly by neutrophils in response to tissue damage and oxidative stress (Kolaczkowska & Kubes, 2013; Margraf et al., 2008). NETs contain histones, neutrophil cytoplasm, and DNA, and are capable of trapping pathogens and damaged tissue (Kolaczkowska & Kubes, 2013; Margraf et al., 2008). The source of cf-DNA from exercise may result in kinetics that are favorable for biochemical athlete monitoring. Previous studies have demonstrated that cf-DNA returns to baseline within two hours after accustomed exercise, in both aerobic and resistance training (Atamaniuk et al., 2010; Atamaniuk et al., 2004). The kinetics of cf-DNA may provide an advantage in athlete monitoring when compared to more commonly used markers. Biochemical markers such as creatine kinase, C-reactive protein, and interleukin-6 require 12 to 72 hours to return to baseline (Barquilha et al., 2011; Breitbach, Tug, & Simon, 2012; Gordon III et al., 2017; Hazuda, Lee, & Young, 1988; Pepys & Hirschfield, 2003; Rogers, Stull, & Apple, 1985). The nature of cf-DNA does not require participants and athletes to cease training while the marker returns to homeostatic values and risk inaccurate representation of the resistance training session of interest.

There is evidence that suggests increases to circulating cf-DNA may be correlated with changes in training volume. Tug et al. (2017) demonstrated greater cf-DNA values in capillary plasma from finger sticks as participants completed additional resistance training exercises. Similar findings have been reported across studies using half and ultra-marathons as the exercise stimulus. Atamaniuk et al. (2004) demonstrated that cf-DNA was significantly elevated immediately after a half marathon and returned to baseline within two hours post. Only when the study was replicated using ultra-marathon participants did cf-DNA remain elevated above baseline beyond 2-hours, while returning to normal levels within 24-hours (Atamaniuk et al., 2008). The current body of research describing the exercise response of cf-DNA is limited, especially in relation to resistance training. Additional research is required using the gold standard real-time PCR to determine the response and sensitivity of cf-DNA after a resistance training session. Further understanding the kinetics of cf-DNA in relation to varying resistance training volume-loads may allow researchers and sport scientists to make decisions about its usefulness in athlete monitoring. Therefore, the purpose of this study was twofold. First, to assess the sensitivity of cf-DNA to different resistance training volume loads, within a heterogenous group of participants. Second, to examine the relationship between cf-DNA and relative strength.

METHODS

Experimental Approach to the Problem

Participants reported to the lab 48 hours before the resistance training session for back squat one repetition maximum (1RM) testing and body mass assessment. The resistance training protocol began with a general and a squat specific warm-up, followed by six sets of ten repetitions of the back squat. Blood was drawn at three different times; immediately before the resistance training session, immediately after the third lifting set, and immediately after the sixth lifting set. Blood was centrifuged to isolate plasma and then frozen. cf-DNA was isolated from plasma and changes in cf-DNA were assessed between T1, T2 and T3.

Participants

Thirty healthy participants were recruited for this study, 15 males and 15 females. Participant body mass and strength characteristics are listed in Table 1. Participants had a minimum of 6 months of resistance training. The NSCA Health/ Medical Questionnaire was used to assess participant health status. No health concerns were reported that would make participants ineligible. Participants were notified of the potential risks and benefits of the study and were provided and signed an informed consent document. This study was approved by the East Tennessee State University Institutional Review Board.

Table 4.1

Participant Descriptive Characteristics (Mean ± SD)

	Body mass (kg)	One Repetition Maximum (kg)	Relative Strength (1RM/BM)
Males	88.44 ± 12.2	153.5 ± 34.0	1.7 ± 0.3
Females	65.2 ± 9.7	71.2 ± 14.6	1.1 ± 0.2

One Repetition Back Squat Assessment

Forty-eight hours prior to completing the resistance training session, participants reported to the lab to obtain their body mass and back squat 1RM. Participants were asked to refrain from additional exercise 24 hours prior through the remainder of the protocol. Body mass was measured using a digital scale (Tanita B.F. 350, Tanita Corp. of America, Inc., Arlington Heights, IL). Participants were asked to estimate their 1RM for the warm-ups and initial attempt. A previously published protocol was followed to obtain true 1RMs (McBride et al., 2009). A generalized warm-up was completed, followed by 10 repetitions of the back squat at 30%, 5 reps at 60%, 3 reps at 70%, and 1 rep at 90% of each participants estimated 1RM (McBride et al., 2009). A maximum of four 1RM attempts were allotted with 3-5 minutes of rest between attempts (McBride et al., 2009). Relative strength was calculated as 1RM kg/ body mass kg (Schilling et al., 2010).

Resistance Training Session

The testing session began with pre-session blood draws followed by a generalized warm up and a squat specific warm-up, consisting of two sets of 10 repetitions; the first at 50% and the second at 75% of the weight prescribed for the 6x10 (Willardson, Kattenbraker, Khairallah, & Fontana, 2010). Two minutes of rest was allotted between the warm-up sets (Willardson et al., 2010). Three sets of ten repetitions of the back squat were completed at 60% of the participant's 1RM, with two three minutes of rest between sets. Blood was drawn after the third set, followed by three additional sets of ten repetitions of the back squat with two to three minutes of rest between sets. A final blood draw was taken after the sixth set.

Biochemistry

Blood was centrifuged (SmithKline Beecham VanGuard 6000 centrifuge, Haddonfield, New Jersey) for fifteen minutes at 6,000rpms and the plasma stored in a -86°C (Thermo Fisher Scientific, Waltham, MA). The plasma was thawed and cf-DNA isolated using QIAamp MinElute ccfDNA Midi Kits (Qiagen, Germany) (Ok et al., 2020). Cf-DNA was analyzed via qPCR (Bio-Rad CFX96 Touch™ Real-Time PCR detection system, München, Germany) (Tug, et al., 2017) with primers copying the RNase-P gene (Malik, Shahni, Rodriguez-de-Ledesma, Laftah, & Cunningham, 2011; Marcuello et al., 2005). cf-DNA is expressed as concentrations in plasma and as percent change (%Δ).

Statistical Analysis

Data were analyzed with SPSS (Version 25.0. Armonk, NY: IBM Corp). Descriptive statistics in the form of means and standard deviations were calculated for all variables. A repeated measures ANOVA was used to assess changes in cf-DNA between T1, T2 and T3. A Greenhouse-Geisser adjustment was used when Mauchly's test of sphericity found the assumption of sphericity to be violated ($p \leq 0.05$). Post-hoc analysis was completed using the Bonferroni correction. Cohen's *d* effect sizes (ES) were reported for all comparisons and classified as <0.1 = trivial, 0.1 to 0.29 = small, 0.3 to 0.49 = moderate, and >0.5 = large (Chen, Cohen, & Chen, 2010).

Two simple linear regressions models were created to assess the ability of relative strength to predict %Δ cf-DNA (T1 to T2 and T1 to T3). Residual plots were used to determine homoscedasticity and normality of the residuals. Cook's distance was calculated to assist in the identification of outliers and Winsorization was used to adjust for any outliers identified using a Cook's distance greater than 1 (Shete et al., 2004).

RESULTS

The means, standard deviations, and relative changes in cf-DNA can be found in Table 4.2. cf-DNA significantly increased from T1 to T2 ($p < 0.01$, ES = 0.96) and T1 to T3 ($p < 0.01$, ES = 1.06). There was no significant difference in cf-DNA when comparing time points T2 and T3 ($p = 1.00$, ES = 0.10).

Table 4.2

Results from Repeated Measures ANOVA

	T1	T2	T3
cf-DNA (ng/µl)	407.72 ± 320.83	1244.6 ± 875.83*	1331.15 ± 1141.66*
%Δ cf-DNA	Baseline	287.68 ± 468.51	280.16 ± 410.47

Significant differences from T1 are denoted as *.

Relative strength was unable to significantly predict %Δ cf-DNA from T1 to T2 ($F_{(1, 28)}$ = 3.79; p = 0.06; R^2 = 0.12) (Figure 4.1). The linear regression model calculated to predict cf-DNA from relative strength (T1 to T3) was significant ($F_{(1, 28)}$ = 4.91; p = 0.04; R^2 = 0.15) (Figure 4.2). A single value was found to have a Cook's distance of 1.15 and was Winsorized.

Figure 4.1

Linear regression model using Relative Strength to predict %Δcf-DNA T1 to T2

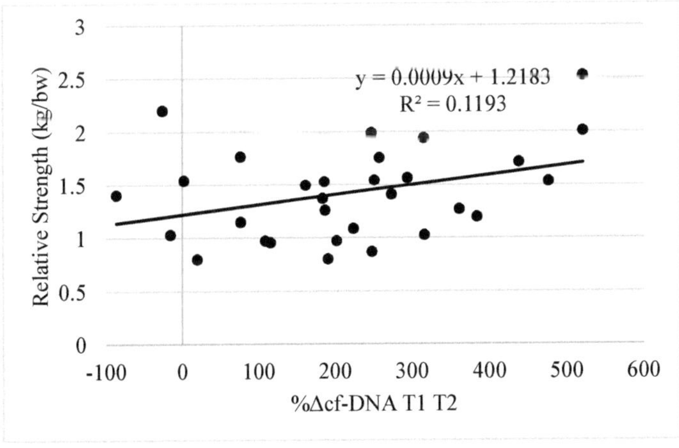

Figure 4.2

Linear regression model using Relative Strength to predict %Δcf-DNA T1 to T3

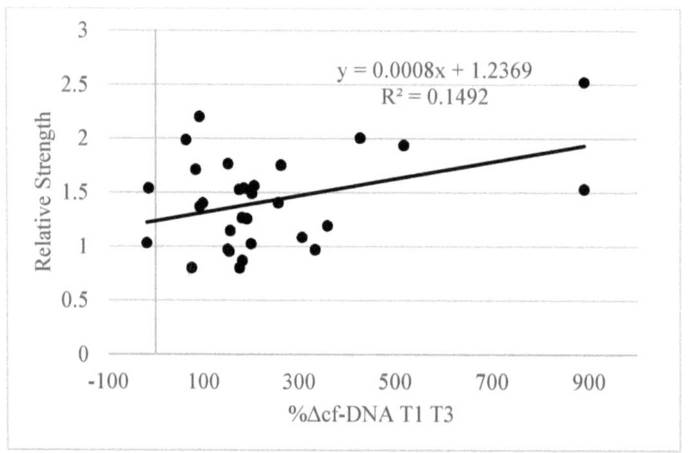

DISCUSSION

The primary purpose of this investigating was to assess the sensitivity of cf-DNA to different resistance training volume loads, within a heterogenous group of participants. The secondary purpose was to explore possible predictors to changes in %Δ cf-DNA, including relative strength. The primary purpose was examined using 6 sets of 10 repetitions of the back-squat exercise with blood draws at 3 different time points. Results were assessed using a repeated measures ANOVA to determine any significant differences between %Δ cf-DNA from T1 to T2, T1 to T3, and T2 to T3. Two simple linear regressions were created to explore the secondary purposes. Relative strength was unable to significantly predict %Δ cf-DNA from T1 to T3, but significantly predicted %Δ cf-DNA from T1 to T3 (Figure 1 and Figure 2).

To the researcher's knowledge, the current study is the first to examine $\%\Delta$cf-DNA at varying volume-loads using the same resistance training exercise. Tug et al. (2017) conducted a study that successfully demonstrated significant intra-session increases to baseline cf-DNA when compared to values following exercises 13 through 16 in a 16-exercise resistance training protocol. The protocol implemented 5 repetitions of each exercise at 60% of the participants reported 1RM (Tug et al., 2017). The high number of resistance exercises required to significantly elevate cf-DNA in the study conducted by Tug et al. may be due to participants only completing 1 set of 5 repetitions of each exercise. The protocol in the current study required participants complete 3 sets of 10 repetitions between blood draws, resulting in a much higher volume-load between draws compared to the protocol used by Tug et al.

Results from the repeated measures ANOVA suggest that cf-DNA will significantly increase in response to higher resistance training volume-loads. The finding that cf-DNA did not significantly increase, nor have a meaningful effect size, from T2 to T3 was unexpected. A possible explanation for this finding is the concept of neutrophil overshoot. Neutrophil overshoot has been traditionally applied to trauma rather than exercise and occurs when an excessive number of neutrophils are recruited following a traumatic stimulus, resulting in an "inflammatory second hit" (Margraf, Ley, & Zarbock, 2019; Margraf et al., 2008). A less severe version of this overshoot may manifest from a high resistance training volume-load such as 3x10 back-squats, decreasing additional neutrophil signaling after subsequent sets. A catecholamine overshoot in response to exercise may be evidence that supports this idea, as catecholamines aid in neutrophil mobilization through hemodynamics (Suzuki et al., 1996). Further evidence is required to explore the possibility of this suggestion.

The two linear regression models elicited results that failed to support the initial hypothesis of a negative relationship between relative strength and %Δcf-DNA. The model created for the predictive capabilities of relative strength to %Δcf-DNA from T1 to T2 trended towards significance in the positive direction , while the second model, T1 to T3, was significant in the positive direction and included a weak effect size ($R^2 = 0.15$). These findings could be due to a tendency of leukocyte response to increase as training status increases (McFarlin, Flynn, Phillips, Stewart, & Timmerman, 2005), although the literature on this is insufficient. A more likely second option is that individuals with greater relative strength lifted a greater resistance training volume-load. Immune response to exercise is influenced by training volume-load (Coutts, Reaburn, Piva, & Rowsell, 2007; Niemelä, Kangastupa, Niemelä, Bloigu, & Juvonen, 2016), suggesting neutrophil response, and therefor %Δcf-DNA, would be similar between participants of varying relative strength levels if the absolute intensity remained constant.

The tendency of cf-DNA to increase significantly immediately after various training modalities has been shown consistently throughout the literature (Atamaniuk et al., 2008; Atamaniuk et al., 2010; Haller et al., 2018; Tug et al., 2017). The growing body evidence that supports the sensitivity of cf-DNA in response to exercise training makes it promising tool in athlete monitoring. Commonly measured biochemical markers seem to lack of consistency seen with cf-DNA in response to training. As evidence increases suggesting cf-DNA is a useful biochemical marker in athlete monitoring, it is important for researchers to discover approaches to data collection that decrease the cost and time required. Tug et al (2017) demonstrated that plasma cf-DNA collected from capillary finger sticks into EDTA vacutainers, rather than venipuncture, still increased significantly between varying volume-loads. Researchers should examine the method completed by Tug et al. compared to venous blood draws to elicit any important differences between outcomes of the two protocols. A second factor that would increase the practicality of athlete monitoring using cf-DNA is the method of quantification. To the researcher's knowledge, there are currently no studies conducted that use multiple methods of quantifying cf-DNA in response to exercise. Holdenrieder et al. (2005) explored the relationship between the standard qPCR and enzyme-linked immunosorbent assay (ELISA) methods of quantifying cf-DNA from patients with pancreatic cancer in response to radiochemotherapy, in serum and plasma. The researcher's results demonstrated significant correlations in cf-DNA values between the two methods when using both serum and plasma, with a strong correlation found with serum but only a moderate correlation with plasma (Holdenrieder et al., 2005). The relationship between qPCR and ELISA should be further investigated through exercise training studies, as ELISA is more affordable and accessible option for athlete monitoring.

CONCLUSION

The present study demonstrated that cf-DNA significantly increases immediately following a high-volume resistance training protocol. The increase of cf-DNA is partially dependent on the relative strength of an individual. The finding that greater cf-DNA values can be expected in individuals possessing greater relative strength should be considered when using the biochemical marker in an athlete monitoring program. Further research is required to validate more cost effective and convenient methods of quantifying the response of cf-DNA to exercise training. The growing body of research continues to support the possibility that cf-DNA may a superior choice when using biochemical markers in athlete monitoring, due to the sensitivity and kinetics of cf-DNA.

CHAPTER 5. SUMMARY AND FUTURE DIRECTIONS

The primary purposes of this dissertation were to explore relationship between cf-DNA, CK, CRP, vertical jump testing, and delayed onset muscle soreness in response to a high-volume resistance training protocol, and to assess the sensitivity of cf-DNA to different resistance training volume loads. The secondary purpose was to examine the relationship between cf-DNA and relative strength. These purposes were accomplished by conducting two research protocols. The first study was designed to explore the relationship of cf-DNA to CRP and CK in response to a high-volume resistance training protocol. Results were largely inconclusive with only the correlation between %Δ DOMS 48hr and %Δ CRP 48hr for the non-supplement group found to be significant ($r = 0.71$; $p = 0.02$). The second study examined the sensitivity of cf-DNA in response to different resistance training volume loads. Results from this study demonstrated significantly elevated cf-DNA following 3 sets ($p < 0.01$, ES = 0.96) and 6 sets ($p < 0.01$, ES = 1.06) of the back squat compared to baseline, with no differences between 3 and 6 sets ($p = 1.00$, ES = 0.10). A linear regression model created to predict %Δ cf-DNA using relative strength was significant for T1 to T3 ($F(1, 28) = 4.91$; $p = 0.04$; $R^2 = 0.15$), suggesting less cf-DNA will be released after resistance training in individuals with lower relative strength, when relative intensity is constant.

Additional research is required to identify any relationships between cf-DNA and DOMS. The practicality of using cf-DNA as an athlete monitoring tool requires greater exploration. Protocols examining the validity of measuring cf-DNA from whole blood finger sticks in athletes could make the use of cf-DNA possible outside of a laboratory.